The authors are professional chimney sweeps residing in Stowe, VT.

Chimney and Stove Cleaning

by
Christopher Curtis
&
Donald Post

Stove and chimney cleaning traditionally have been thought of as despicable tasks relegated to those unfortunate individuals doomed to the role of chimney sweep. However, almost anyone who has a stove and chimney can clean them, using a little elbow grease, and a lot of soap. Slightly soiled hands are, after all, a small price to pay in return for the millions of BTUs your stove has provided you, so take heart, and when your hands are black . . . smile.

Many people wonder why a stove ever needs cleaning, so before we get to the HOW, let's look at the WHY. When wood burns, its combustion is never complete. Wood smoke inevitably contains unburned gases and vapors which condense on contact with any surface whose temperature is below their dew point. This occurs when a cold stove is being fired or when smoke contacts the cold stovepipe exposed to outside temperatures. This condensate is called creosote and is flammable.

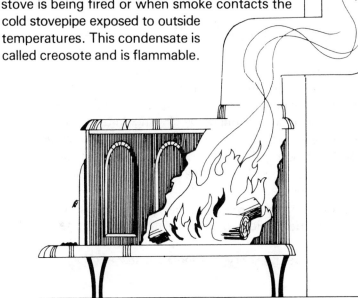

Changed by Heat

Raw creosote condenses from flue gas as a liquid. It soon dries to a shiny, hard coating. Creosote in this form is flammable, but does not cause chimney fires. Chimney fires usually start after creosote has been pyrolyzed, or changed chemically by heat to form a soft, black, crusty deposit on the inside of the chimney. If coal is burned in the stove, creosote will not form, but a flammable substance, coal tar, is deposited in the stack. Periodic cleanings are required.

When a chimney fire occurs, the building is jeopardized in several ways: stovepipe can glow cherry red, threatening anything that will burn, especially flammable materials near where the pipe penetrates the ceiling or roof. Burning debris can shoot out of the top of the chimney, landing on the roof or nearby bushes.

The metal walls of the stovepipe can burn through if the fire is prolonged or intense, especially if there is excessive corrosion. Creosote is acidic, and if combined with water attacks steel and mortar. Condensation of water can be avoided if a stack temperature of at least 150°F. is maintained.

Another Hazard

A less obvious but equally insidious hazard is presented by the chimney fire that ignites suddenly, consumes all the available oxygen almost instantly and extinguishes itself, allowing oxygen to rush in and be ignited by glowing creosote. This on/off cycle is repeated two or three times a second, resulting in a violent vibration in the stovepipe which may shake itself loose, showering the bewildered inhabitants with red hot steel and flaming creosote.

Chimney fires in prefabricated chimneys may warp or burn on the inside with no signs of damage on the outside.

Ceramic and masonry flue liners are susceptible to cracking

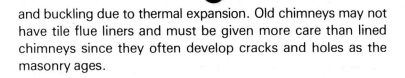

and buckling due to thermal expansion. Old chimneys may not have tile flue liners and must be given more care than lined chimneys since they often develop cracks and holes as the masonry ages.

Settle and Crack

Masonry chimneys are heavy by nature, and some settling is inevitable. Many times cracks will result from settling. Chimney fires in old chimneys are particularly dangerous because of these holes as well as their method of construction, which involved single brick walls, sometimes even single bricks laid edgewise. Cracks and holes can allow sparks and embers to ignite dry, aged wood adjacent to the chimney. Thin walls of older chimneys can transmit the intense heat of a chimney fire to combustibles outside the chimney.

If holes and cracks in an old chimney cannot be repaired satisfactorily, it may be possible to install stovepipe inside the chimney. This is done by lowering stovepipe into the chimney, attaching section after section, until the desired depth is reached. Connections to appliances are made, and the whole assembly is hung from the chimney top.

Efficiency Reduced

Heating efficiency is also a reason to keep your stove and stovepipe clean.

(A brief digression may be appropriate here for definition of terms. A stove is a stove is a heating stove is a cooking stove. Stovepipe is a section of thin-walled, usually uninsulated pipe leading from the stove to the chimney or stack. A stack or chimney is a heavier continuation of the smoke venting system that is usually predominantly vertical, that can be insulated stovepipe or a prefabricated chimney, or a conventional brick and mortar chimney.)

Soot and creosote are excellent insulators. Their thermal conductivity varies, but it can be similar to that of glass wool

3

insulation. Since the surface area of the stovepipe often exceeds the heating area of the stove, significant portions of the total heat output can be extracted from the stovepipe. Small accumulations of soot and creosote radically reduce the heating value of stovepipe. Likewise, soot accumulations within stoves, particularly baffled models, reduce heating efficiency. For maximum heating efficiency, your stove and stovepipe should be cleaned often.

Less Smoke

Aside from the safety and efficiency benefits of chimney cleaning, there is the convenience afforded by an unrestricted flue. Smoking when the flue door is open and backdrafting are sometimes indications of a partially blocked chimney. A clean chimney reduces the tendency to smoke, making loading the stove easier and more convenient.

Look inside the stovepipe and chimney to determine whether they should be cleaned. It is wise to inspect both the stove end and the top of the chimney, as a minor deposit at the stove exit may indicate a heavier deposit further from the stove where flue gas condensation is heavier. Condensation is naturally heavier where the stack is cold. If the chimney is outside the heated portion of the house, buildup will be faster than if it is inside the house. Insulated prefabricated chimneys support a higher internal temperature, minimizing creosote buildup. Often the greatest buildup occurs in the last few feet of chimney, the part exposed to the outside temperatures above the roof. If the buildup of soot and creosote is ¼-inch or thicker, clean the chimney.

Rate Varies

The rate that the creosote forms has many variables, but the most important are the type of wood burned and the stove tightness. An airtight stove can generate chimney fire material in only three or four weeks of heavy burning. A stove that is not particularly airtight could go an entire season with good wood before cleaning is necessary. In any event, frequent inspection is recommended, particularly with a new stove whose operating characteristics are not fully understood.

Frequent inspection is needed when a new pile or different truckload of wood is first being burned. Different vintages and types of wood yield different amounts of creosote. A new pile may contain wood that will build creosote much faster than the wood you've grown accustomed to. Inspect the pipe often after installation of a new stove, or when starting on a new batch of wood. Chimney cleaning is recommended at least at the end of each heating season or whenever inspection shows more than a ¼-inch thick accumulation.

Prepare for the Task

When the need has been determined, the next thing to do is prepare mentally for the task. As in most chores, the notion of ease and expediency must be kept foremost in one's mind. Convince yourself that you can have as much fun cleaning your chimney as the sweeps in Mary Poppins. Once you have mentally prepared, assemble the necessary tools and begin by cleaning the stove. The stove does not need cleaning as a fire precaution, but to improve heat transfer efficiency. Cook stoves should be cleaned periodically to maintain high cooking surface temperatures.

Tools You'll Need

Although all the tools below are not mandatory, the more you gather before starting, the smaller the chances of problems arising. Beg, borrow, or steal a:

- drop cloth
- trouble light
- old pair of leather work gloves
- hand wire brush
- hand scraper or stiff putty knife
- hammer and screwdriver
- ash shovel
- metal bucket
- vacuum cleaner
- dust pan
- whisk broom

A can of furnace cement and an adjustable wrench are good standbys if you must disassemble any parts of the stove.

Before beginning, be sure the stove is stone cold. Working on and in even a warm stove is possible, but highly undesirable.

If it is warm, spread the drop cloth on the floor in front of the fuel door, open the door, and shovel the ashes and embers into the metal bucket, take the bucket outside, and wait till the stove cools. When it is cool, remove the remaining ashes either by the ash drawer if the stove is so equipped, or by shoveling if necessary.

If you have no grate in the stove there is probably sand on the floor of the firebox. Take care not to shovel it out with the ashes. Ideally everything should be removed, but a sand lining for the bottom of the firebox is required, and if you are working in the winter, finding sand may prove difficult. Rather than risk it, leave a few inches of ashes and sand in the bottom of your grateless stove.

If it is necessary to add sand, do not add sand intended for road use, as it may contain salt. Salt and iron do not complement each other. Save whatever ashes you remove for use in the garden.

If the stove is small, consider taking it outside. This may prove easier and cleaner in the end.

Use the trouble light to illuminate the interior of the stove and peer in. Once you have an idea what the inside looks like, use your gloved hand (the gloves should have been on since you touched the stove to see if it was cold) to wire brush and/or scrape every accessible square inch inside the stove. If your firebox is box-shaped with the smoke port exiting directly into the stovepipe you're almost done.

Clean Baffle System

If your stove has a baffle system or a secondary combustion chamber, you must clean this area. Many stoves have push-out panels to open the secondary chamber. These panels probably were sealed with furnace cement during assembly and may be difficult to remove. If you must resort to impact persuasion (with your hammer) do not hit the metal directly. Use a block of wood to absorb some of the shock, as cast iron does not accept criticism gracefully.

If your stove is not equipped with push-out panels but does have inaccessible baffling, it is probably bolted or screwed together. Locate the likely fasteners and unscrew them. If the panel will not move, assume furnace cement has been used, and try the hammer and block for some friendly persuasion.

After the panel has been removed, wire brush or scrape the interior of the chamber. The vacuum cleaner may be used to remove all the loose soot and creosote from the chamber. Scrape off whatever furnace cement remains on the contact surfaces. Inspect all parts for cracks, corrosion, and heat warpage before applying new cement and reassembling the stove.

Fingers Work Best

A putty knife works to apply the cement, but nothing works as well as your fingers. Follow the directions on the can, applying just enough to seal the joint. Let the cement dry twelve to twenty-four hours before firing the stove.

The final step is to put the light into the stove and make a thorough inspection, looking for cracks. Cracks in the stove walls sometimes occur due to thermal expansion. These cracks as well as all the joints should be sealed with furnace or refractory cement. Generally, if the plates on either side of the crack are mechanically stable, cement can be successfully applied. If the plates can be moved, have the crack welded.

Inspect Door Seal

Inspect the door seal to make certain there is a tight fit of the metal closure or asbestos gasket type seal. If the seal is incomplete the stove cannot be tightly closed; order a new seal, but wait until it arrives before removing the old one. Replace the sand in the bottom of the stove if needed, and the stove is done. If you're going to clean the chimney, leave the drop cloth in place and proceed; otherwise pick up the tools and drop cloth and go wash up. Then return to vacuum whatever little soot remains.

Preparing for Storage

If the stove will be stored after the cleaning, remove all ash and sand to prevent corrosion. Store the stove in a dry place. Humid summer weather can promote corrosion and rust.

A summer coat of stove black can be applied, but is generally unnecessary. If the stove has rust spots or spots of shiny metal, use stove black. It will inhibit further rusting, where rust has begun. In the case of shiny spots, it is desirable to use stove black for two reasons. The first is cosmetic. A uniformly flat black stove looks like new. The second reason is function. Shiny bare metal's emissivity is much lower than the same metal painted flat black. That means that in terms of radiation output, shiny spots are cooler than black areas next to them. Shiny spots are also great spots for rust to start.

Cook Stove Cleaning

Cleaning a cook stove is similar to cleaning a heating stove, with a few exceptions. Heat transfer efficiency is of great importance, so keep your cook stove clean. Again, do not attempt to clean the stove while it is hot or warm. Collect the same tools as for the heating stove and lay out the drop cloth. Remove all the parts that can be lifted from the cooking surface and, with gloves on, wire brush them. Shovel out the ashes or remove the ash drawer. Save those ashes for the garden, as they may be used to decrease the soil acidity and add phosphorous and potash to it.

Once you have laid open the bowels of the stove, methodically brush or scrape the entire inside. Determine the route of smoke passage from the firebox to the stovepipe and scrub the entire passageway.

The oven area needs special attention. Baking in a wood-fired stove is an art, requiring slow, even heat. Many a wood-baked cake has been burned on one side while still batter on the other. Take great care to clean the heating surfaces of the oven. Some stoves are equipped with special ports for this purpose, others require more patience. Either way, if you bake, keep the oven area clean. Remove the soot and creosote and replace the lift-out pieces.

Cleaning Water Jacket

If your stove has a water jacket or coil that is in use, occasional cleaning of the *water side* is recommended. Frequency of cleaning depends on the mineral content of the water, the volume of water that passes through the jacket and whether a water softener is used. Semiannual cleaning should suffice unless you have excessive minerals in the water.

We must digress here long enough to say that if your stove is equipped to heat water, by all means make use of this provision. Heating water is expensive energywise, and for a small investment, you can at least preheat the water entering your hot water heater and benefit by lowering your gas or electric bill and increasing your supply of hot water.

How to Clean Coil

Cleaning of the coil is facilitated by a strong solvent that removes scale and muck from the inside of the tubing. In the past such strong corrosives as sulphuric and hydrochloric acids were used, but they were dangerous to handle and dispose of, and corroded the metal.

The best solution to use is a solvent that is not acidic until heated and is biodegradable. It is probably not carried by your local stove store, but it's worth shopping for. Try a commercial boiler supply house or a boiler maintenance service. Don't let some slippery-tongued boiler salesman sell you acid; specify the biodegradable solvent.

You will also need a short piece of flexible tubing with a fitting that matches the upper connection of the coil, two buckets, and an end plug that matches the lower fitting of the coil. These fittings are outside the stove.

When you have collected these things, begin by closing the valves that lead to and from the domestic water supply, then uncoupling the top fitting of the coil and attaching the section of flexible hose. Fill the hose with water, and put the end in a bucket of the biodegradable solution. Raise the bucket high enough to set up a siphon and disconnect the lower coupling. You want the water to drain out of the coil into the second bucket, sucking the solution into the coil.

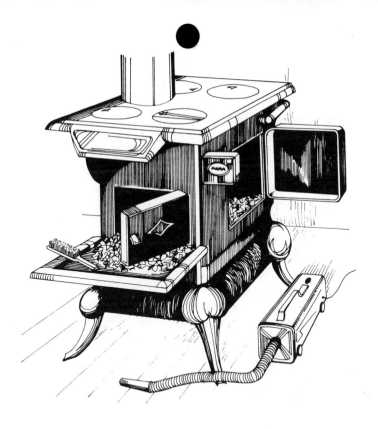

When the coil is filled with solution, seal the lower coupling with the end plug. Light a fire in the stove. The fire is necessary because the solution does not act as a cleaning agent until heated. Maintain a fire just hot enough to boil the solution for four to six hours, remembering that expansion and boiling will push the now acidic solution out of the upper coupling and back into the bucket. If you use plastic pipe or hose, *beware of overheating* the plastic. If the plastic gets too soft, wrap a cold wet towel around the hose and connection.

At the end of the boiling period, remove the plug from the lower fitting and drain the solution, replacing with fresh water via the siphon technique. Heat this water to boiling and redrain, replacing with fresh water. Continue flushing with *cold water* until you feel comfortable reconnecting the domestic water (you're actually safe after the second flush). Despite the biodegradable claim, take the used solution outside and dump it where it can't contaminate open water.

Cleaning the Stovepipe

Cleaning the stovepipe is the next task. The stovepipe leads from the stove to the chimney or stack. Again, how frequently it should be cleaned can be determined by inspection. Stovepipe that carries low velocity, high concentration, relatively cool flue gas from a high efficiency stove may need brushing out every other week. By the same token, stovepipe carrying high velocity, hot flue gas could go an entire season before needing cleaning.

To clean the stovepipe, it is best to have the proper size flue brush, a cylindrical brush, designed specifically for this use. However, I have found that diligence and a toilet brush also will work in a pinch. If you intend to clean your chimney on a regular basis yourself, invest in a flue brush. Flue and chimney brushes come in a wide variety of shapes and sizes. Be sure you know your size requirements before shopping for the brush or brushes you need. If you need only

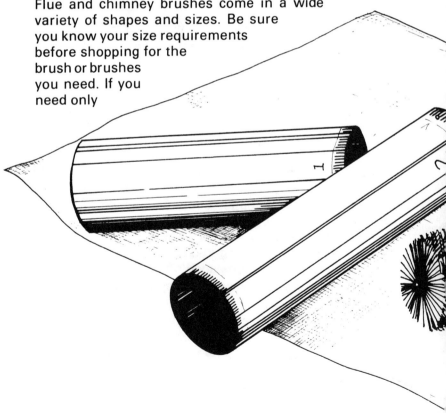

a round brush to fit your round stack, buy the same diameter brush as you have stack. If you need a square or rectangular brush, specify both dimensions. Flue brushes are available in both plastic and steel. Spend the extra cash for steel. They do a better job in less time. These brushes are available from fireplace and stove stores.

Take It Outdoors

The stovepipe sometimes can be cleaned in place, but it's usually easier to take the whole thing outside. To do this, disassemble the pipe into sections that can be conveniently carried outside, but first make a scratch at all the joints that will be taken apart and on all articulations of swivel type elbows. If there will be many pieces, number the sections. When you're trying to rematch screw holes and maintain visual plumb during reassembly, the scratches and numbers will pay off.

The pipe must be disassembled to pieces small enough to avoid bumping anything on the way out, as this inevitably causes soot to fall on a white carpet. Try to carry the sections with

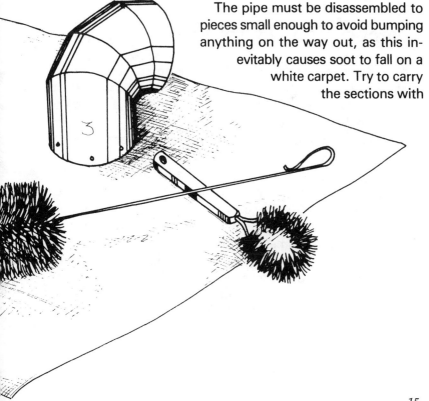

the bends or elbows opening end up to further reduce the possibility of soot fallout. When all of the pieces have completed the treacherous cross-carpet journey, and are safely outside, further disassembly may be needed. The pipe should be disassembled to sections that have no more than one bend to facilitate easy cleaning.

Use the flue brush to scrub the creosote off the surface. Don't expect the pipe to shine; just get it to the point where no thickness of deposit is discernible. Don't be afraid to realign the toilet brush to suit its new use. The wire that holds the brush in shape is malleable and easily lends itself to "design corrections." Brush out the entire length of stovepipe.

If the stovepipe has a downdraft equalizer or a heat reclaimer, it will probably have to be disassembled. Scratch the joint and take it apart. Brush out the equalizer if necessary. Most heat reclaimers have a cleaning plate that operates while the unit is in place. If yours does not, scrub the heat transfer pipes until they shine.

Damper Can Be Problem

The damper in the stovepipe is another potential problem. If this must be removed, reach inside the pipe with one hand and hold the damper plate. With the other hand, push the damper handle in, toward the pipe, and twist it a quarter-turn. This will disengage the handle and shaft and facilitate removal. Simply pull the handle straight out of the hole. Some finesse may be needed, as the shaft has a bend that must come through the hole. Simple rocking of the shaft will usually free it. Note the position of the spring for reassembly. Some dampers are welded in and cannot be removed. If yours is that type, do not waste time attempting to remove it. Brush it in place. Clean everything and inspect for corrosion. Remember, water from rain and condensation combines with creosote to form acid that is corrosive to steel. If corrosion is excessive, replace the pipe.

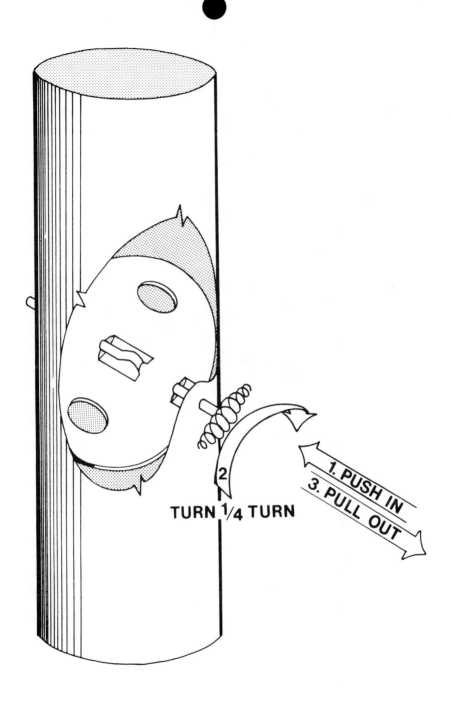

Need New Stovepipe?

If new stovepipe is required, bear in mind that stovepipe can contribute significant portions of heat given the chance. Chrome or bright galvanized stovepipe are bad choices since the infrared emissivity of shiny bare metal is low. Blue-oxide or stoveblackened pipe are better for this reason. Buy the heaviest gauge you can find, especially if you are one of the non-purists who occasionally burns trash in your stove. Smoke from trash corrodes stovepipe fast. Heavy gauge is safer and will last longer.

   Replace the damper and reassemble the pipe to the carrying stage, aligning scratch marks and making certain that each joint has at least three screws. Two screws allow side deflection, but three or more insure relative rigidity. Leave the sections outside until the stack has been cleaned.

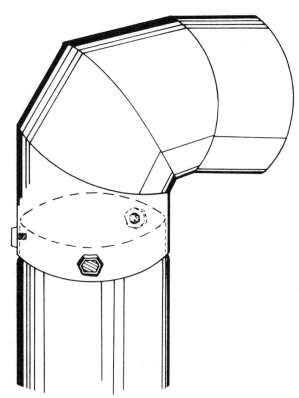

Cleaning the Chimney

The chimney or stack is the most difficult part of the system to clean, but at the same time is the most essential. This is where most of the flue gas condensation and resulting creosote occurs. Condensation of flue gas naturally occurs more readily on cold surfaces than warm, so if the stack is outside the house, build-up is likely to be faster than if it is inside the heated portion of the building. Sometimes creosote deposits are heaviest near the point of smoke entry. This is a good place to inspect and determine the need for cleaning. Sometimes it is possible to clean only the stack without having to remove the stovepipe. If this is the case in your set-up, clean the chimney more often than the stove and stovepipe. How often can only be determined by inspection, so inspect often.

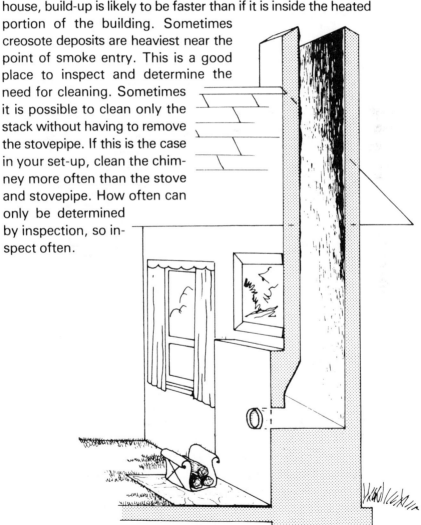

Two Types

There are two basic types of stacks: prefabricated chimneys and masonry chimneys. The prefabricated chimneys are easier to clean. Most prefabricated chimneys are round on the inside, making the round flue brushes of great use.

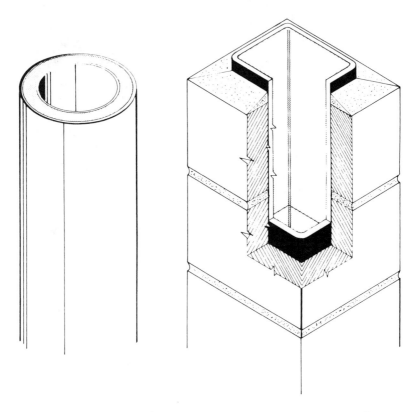

With a proper size flue brush and a few extension handles, one can clean a chimney without going onto the roof. Lay the drop cloth under the stack opening and simply push the flue brush up into the flue. Continue to add extension handles until the brush is at the top of the chimney. A rag held around the handle and over the bottom opening will keep the creosote contained. The brush and handles can be made to turn a corner if the only opening to the stack is on its side. Patience is necessary, but this orientation will work fine. A few up and down strokes should brush out all the dangerous creosote.

If you use the modified toilet brush method, a plumber's snake serves as a serviceable handle for it. Tape it on, and treat it like a lousy flue brush. When you're satisfied that the inside is spotless, remove the brush, shovel out whatever debris remains, replace the cover, and you're done.

Perhaps your system isn't that easy, and you'll have to clean from the roof. If this is the case, leave the stovepipe in place, so that the creosote will fall into the stove instead of into your living room. The round brush method is easier, but if you clean from the roof, you don't need the extension handles. These can be replaced with a weight and a rope. Find a rope long enough to reach the length of the chimney, and strong enough to support the weight. A flue brush and a weight stuck in a chimney because of a broken rope is a chimney's nightmare, so don't use weak rope. The weight should be heavy enough to drag the brush down the chimney — twenty pounds is usually enough. A twenty-pound length of heavy chain is a good weight, or you can make one by filling a can with rocks, metal, or concrete. It is also good to take a screwdriver, the wire brush, and a couple of adjustable wrenches on the roof with you.

Getting on the Roof

Common sense will dictate the best way of getting on the rooftop, but there are some established guidelines. If possible, put the ladder against a gable rather than against the side of the roof. Place the ladder so that it makes about a *four to one slope* with both legs firmly planted. If possible, have someone steady the ladder for you. If the stack is near the peak, you're all set, but if it punches through in the center of one side, you'll need another ladder and a ridge hook. Put the first ladder against the side to enable easy placement and use of the second ladder.

Once on the rooftop, you may have to remove a chimney cap or bonnet. It may be secured with a nut and bolt, or some screws, so arrive equipped to deal with either arrangement. Some tops come off with a one-eighth counterclockwise twist. When the cap is off, brush it, then lower the weight with the flue brush attached into the flue. A few trips up and down the chimney should have it sparkling clean. Remove the rope, brush, and weight, replace the cap and you are done.

Masonry Chimneys

Masonry chimneys are slightly more difficult, but the technique is basically the same. Flue brushes are available to fit most tile liners. If you purchase a rectangular brush for the flue, and take care of it, it will outlast your splitting mall and spare you the aggravation of cleaning the flue without it. The procedure with the rectangular flue is the same as for a round flue.

Without the brush, we take a giant step backward into the dark ages of chimney sweeping. There are several alternative methods, all of dubious effectiveness. The most popular second-rate method is the "hanging chain" method. This involves hanging a long, heavy chain down the chimney and depending on its weight and flexibility to knock off the creosote while it is swung and shaken from the top of the chimney. This method clearly cannot do the same kind of job the brush does, but more than that, it can break flue tiles and damage the chimney. Chains or sand in a burlap bag on the end of a rope are also methods of doubtful effectiveness. A small evergreen drawn up and down the flue can be tried. I've also heard of, but don't condone, tying a goose by the feet and letting the flapping wings remove the creosote. Nothing really works as well as the flue brushes, so if you aren't going to buy a brush, consider securing the services of a professional chimney sweep.

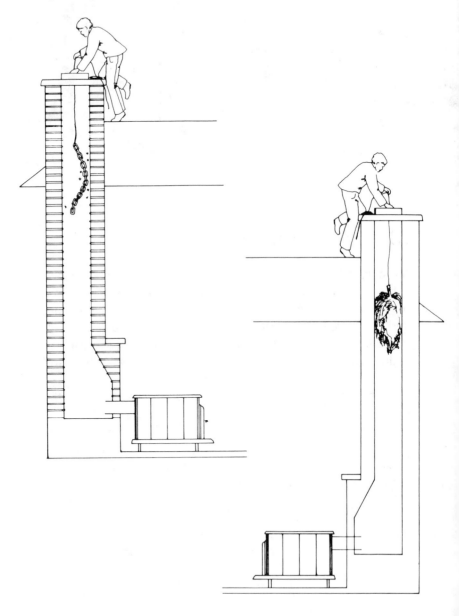

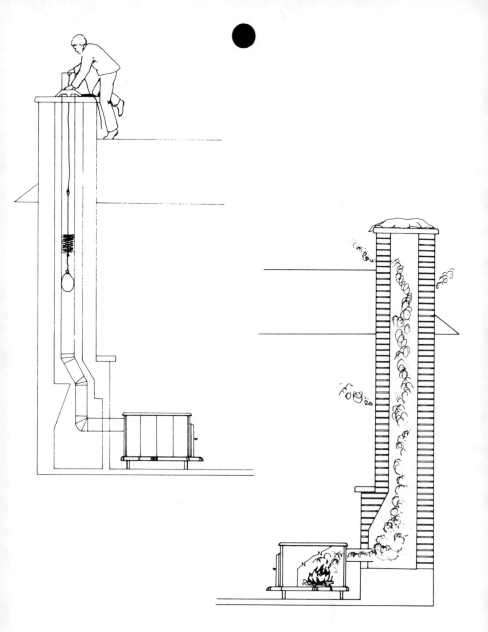

Clean It Out

When the roof portion of the task is complete, you must remove the soot and creosote that have fallen to the bottom. If the chimney has an ash door at the base, shovel out the debris.

If the chimney ends where the stove plugs in, remove the soot there. If you cannot remove the soot, your chimney is doomed to fill up slowly until the point in 1997 when you can shovel a bit out of the hole the stove pipe goes into. Put the debris in a plastic bag and send it to the dump.

When the entire system is reassembled and operational, a smoke test to determine whether there are leaks is a good idea. To do this, solicit the help of a friend. Send him onto the roof with a wet towel, ready to cover the top of the chimney. Go inside and light a very small fire in the stove and add a handful of green leaves or hay or other material that smokes when it burns — you'll do well to stay away from burning rubber and other smelly materials. Have your friend cover the chimney. Now inspect the entire chimney for leaks. If you find leaks, have a mason point up the holes.

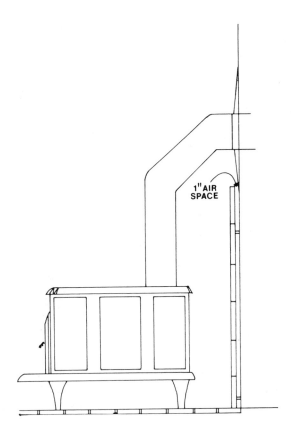

To Avoid a Mess

Within your stove system, there are a number of things that should be done to insure safe, carefree wood burning. A little attention given to these small points will save a mess at least and your house at best.

Creosote sometimes leaks from pipe joints and runs down the outside of the pipe making an unsightly mess. Orient the stovepipe so that the male end of the pipe points down.

Secure it at joints with at least three screws, arranged symetrically to eliminate side deflection.

Heavy gauge stovepipe should be used to avoid the possibility of burnout if a chimney fire occurs. Heavy pipe also resists corrosion.

Avoid long sections of horizontal pipe. Flue gas velocity is low in horizontal pipe, resulting in rapid creosote deposition in these areas.

When horizontal pipe is required, it should have an upward slope of at least 1 in 20.

Maintain a high stack temperature. This will mean some heat loss, but it prevents rapid build-up and condensation of water to form corrosives. Insulating the stack whenever possible will help maintain a good stack temperature.

Burn seasoned hardwood if possible. Green wood and soft wood contain much more pitch and resin that contribute to creosote formation.

When firing a cold stove, allow it to come up to normal temperature quickly, then stop it down.

When convenient, make small additions of small wood often, rather than occasional large loads of large wood pieces.

Avoid closing the stove down with fresh loads.

Keep the stove thirty-six inches from combustibles, or use asbestos mill board or metal sheet to shield walls. Be sure to leave an air space of about an inch for circulation behind this shield.

Use an effective shield on the floor to insure the safety of the floor. Special floor pads are available from your stove store.

Common sense is still the best advice. If you are unsure however, do not hesitate to call an expert.